Cillas Pollicarto da Silva

Soil protein related to glomalin in soils from Ceará

Cillas Pollicarto da Silva

Soil protein related to glomalin in soils from Ceará

And its relationship with degraded areas

ScienciaScripts

Cover image: www.ingimage.com

This book is a translation from the original published under ISBN 978-3-330-75453-9.

Publisher:
Sciencia Scripts
is a trademark of
Dodo Books Indian Ocean Ltd. and OmniScriptum S.R.L publishing group

120 High Road, East Finchley, London, N2 9ED, United Kingdom
Str. Armeneasca 28/1, office 1, Chisinau MD-2012, Republic of Moldova, Europe
Printed at: see last page
ISBN: 978-620-8-24627-3

CONTENTS

SUMMARY

Due to the systematic nature of anthropogenic activities, combined with natural actions, the problem of soil degradation has been increasing and causing a marked process of desertification, which results in loss of biodiversity, erosion and lack of soil fertility. In view of this, the use of tools that indicate the stages of this degradation and recovery is necessary in order to enable appropriate management of these areas. Thus, the study of arbuscular mycorrhizal fungi, as a way of facilitating the recovery of degraded areas, has been increasing, especially due to the production of the soil protein glomalin, which is of great importance for the stability of aggregates and has great potential for use as an indicator of soil quality. Therefore, the aim of this study was to evaluate glomalin-related soil protein concentrations, correlating them with chemical, physical and biological attributes in areas of the municipality of Irauçuba- CE, namely: an area degraded by overgrazing, an area in the process of natural regeneration managed by excluding domestic animals and an area of native forest characterized by shrubby caatinga vegetation. Deformed soil samples were taken at a depth of 0-10 cm in each area and AMF attributes were assessed, such as: total spore density, easily extractable and total glomalin-related soil protein content ($PSRG_{FE}$ and PSRGT); soil chemical attributes (pH, Ca^2 +, Mg^2 +, Al^3 +, P, CO, N) and physical attributes (aggregate stability). The results obtained from the mean tests showed that among the three areas, only the native forest showed significant differences in terms of the two protein fractions (PSRGFE and PSRGT). However, with regard to spore density and aggregate stability, the native forest and exclusion areas had the best average values. The content of PSRG, associated with other edaphic attributes, contributes to discriminating the quality and monitoring of areas with different levels of soil degradation in Irauçuba-CE. The hierarchical clustering analysis showed that three groups were formed, proving the distinction between the native forest area and the areas of overgrazing and exclusion, as well as the similarity of the latter two areas, indicating that more time is needed for significant differences to occur between these areas in terms of their recovery.

Keywords: glycoportein, arbuscular mycorrhizal fungi, degraded area

1 INTRODUCTION

The removal of natural vegetation in the semi-arid regions of northeastern Brazil and the intense movement of the soil combined with long periods of drought promote physical, chemical and biological changes in the system. These changes can lead to marked degradation of the soil, exposing it to the actions of the climate and consequently reducing its productive potential, and can even cause irreversible damage to the environment, compromising the attributes of the soil, especially with regard to the edaphic microbiota and the natural renewal of the region's native species. Revegetation of the soil promotes its recovery and protection through the distribution and incorporation of nutrients, forming a new environment for soil organisms, which are important for the formation and conservation of plant communities.

The activity of edaphic microorganisms plays an important role in the process of plant succession. Among the most influential members of the soil microbiota, helping in the recovery processes of degraded areas, are arbuscular mycorrhizal fungi (AMF), which can collaborate with plants under inhospitable conditions, improving the absorption of nutrients, especially phosphorus. AMF are essential components of ecosystems, contributing to the state of aggregation of eroded soils and therefore reducing the risks of further degradation in areas most susceptible to desertification.

The importance of the diversity and density of AMF species is not only limited to their ability to mobilize nutrients and improve soil aggregation, but they can also be considered as a tool for measuring soil quality. In this context, the production of the glomalin-related soil protein produced by glomalean fungi stands out, not only because it is an important component of the soil's carbon stock, but also because it has great potential as an indicator of the soil's biological quality, an aspect that is still little considered in the assessment of degradation.

Thus, we tested the hypothesis that soils under different degradation conditions show variations in the content of soil protein fractions related to glomalin and these, when associated with other edaphic attributes, reinforce their inclusion as an important quality indicator variable in monitoring soil degradation.

The aim of this study was to evaluate soil protein concentrations related to glomalin, correlating them with chemical, physical and biological attributes in areas of the municipality of Irauçuba- CE, namely: an area degraded by overgrazing, an area managed by excluding domestic animals and finally an area of native forest.

2. LITERATURE REVIEW

2. 1. General aspects of agricultural soil degradation

Degradation is a problem of global dimensions, resulting from various factors involving climatic variations and human activities. Among the processes of soil degradation, accelerated erosion is the most serious form of land degradation worldwide. It is defined as a process of detachment, dragging and deposition of soil particles by an erosive agent. According to Drumond et al. (2004), the result of the cumulative effect of environmental degradation is considered one of the main causes of desertification.

According to Moreira (2004), degradation is a human-induced process or a natural accident that reduces the current and future productive capacity of the ecosystem. Heavily degraded areas can be classified as natural extensions that have lost their ability to recover naturally after undergoing major alterations. According to the Brazilian Association of Technical Standards (ABNT, 1989), soil degradation can be understood as an adverse alteration of the soil's characteristics in relation to its various possible uses, both those established in planning and potential uses, or simply the decrease in soil quality caused by anthropogenic processes. The Brazilian Institute for the Environment and Natural Resources (IBAMA, 1990) states that degradation occurs when native vegetation and fauna are destroyed, removed or expelled; the fertile soil layer is lost, removed or buried, and the quality and flow regime of the water system are altered, along with the socio-economic unfeasibility of the area.

The changes caused to natural ecosystems occur to the extent that they are being replaced by activities aimed at industrial purposes or food production, leading to degradation caused by improper use and management of soils (CENTORION et al., 2001).

One of the main forms of soil degradation is the removal of natural vegetation or its intensive use, actions that promote physical, chemical and biological changes in the system, thus compromising edaphic quality and the natural regeneration process of native species (SILVA et al., 2010). These changes imposed by man on ecosystems in various parts of the world have altered the geographical environment in such a way that the damage caused by these alterations has motivated a large number of researchers to study ways of recovering these altered areas. In this sense, soil is one of the most studied elements of the ecosystem, since its recovery, when degraded, can be slow and not easily reversible (ARAUJO et al., 2008).

Brazil has been suffering from the problem of soil degradation for some time now. This problem began in the colonial period with the Brazilwood cycle and soon after with the sugar, gold and coffee cycles, which led to the formation of extensive deforested and burnt areas (GONÇALVES et al., 2003).

According to Siqueira et al. (2008) there are three strategies that are used to reverse soil degradation, namely: rehabilitation, which is the return of the degraded area to an appropriate biological state; recovery, which means the return of the degraded site to a form and use in accordance with the pre-established plan for land use; restoration, meaning the return to the original state of the ecosystem, a condition that is often difficult to re-establish (MAJER, 1989). All these strategies have revegetation as their main form of action.

There are currently 200,000 km^2 of degraded land in the Brazilian semi-arid region and, when added to the areas where the desertification process is already occurring moderately, the total area affected by the phenomenon reaches approximately 600,000 km^2 , or around 1/3 of the entire northeastern territory (PEREZ-MARIN et al., 2012).

The risk of desertification occurring in Brazil is restricted to the semi-arid region, where its unmistakable characteristics, such as the occurrence of patches of shallow soil, rocky outcrops and maximum dehydration, justified the use of the term Desertification Nuclei (VASCONCELOS, 1982). This problem covers almost all areas of the semi-arid region, turning formerly productive land into unproductive land and thus affecting the agriculture and livestock sectors, increasingly restricting development and the economy, as well as the quality of human life in the affected areas.

In Brazil, there are four regions with low susceptibility to degradation, all of which are located in the semi-arid region of the Brazilian Northeast: Gilbués, Irauçuba, Seridó and Cabrobó.

2. 2 Characterization of the study area

The state of Ceará has a territorial area of approximately 148,000 km^2 and can be considered one of the best representatives of the semi-arid region of the northeast, since around 92% of its territory is subject to adverse climatic conditions. As a result, it has the typical problems of semi-arid areas, especially those susceptible to desertification.

The municipality of Irauçuba/CE has a landscape that reflects peculiar climatic, geological and geomorphological conditions, which makes it drier than other areas in the semi-arid region of Ceará (SOUSA, 2009). The predominant economic activity in this region is extensive livestock farming under conditions of overgrazing, which ends up having a significant impact on these ecosystems due to the removal of part of the plant biomass, causing a reduction in the amount of litter on the soil surface, with negative consequences for the conservation and cycling of nutrients (SALOMON et al., 2004). In these areas, these problems are more intense due to the long periods of drought and the usually high temperatures.

The problem of soil degradation in semi-arid regions is considered serious, justifying the need for more studies related to the evaluation of the physical, chemical and biological aspects of these soils, in an integrated manner, which will thus support the understanding and management of these degraded areas. In the region of the municipality of Irauçuba/CE, where this study was conducted, research is already being carried out to assess the recovery of degraded areas that have been preserved for more than twenty years by monitoring the main chemical, physical and biological attributes, as well as the re-establishment of the fauna and flora characteristic of the site.

According to Brasil (1973), the soil survey of the state of Ceará indicates eight soil associations in the study area which, according to the Brazilian Soil Classification System (EMBRAPA, 2006), are classified as Argissolos, Luvissolos, Neossolos and Planossolos. The Planossolos, however, represent 18.7% of the municipality's soils

(MESQUITA et al., 1981). In the exclusion areas where the soil samples were collected for this project, associations of Planossolos and Luvissolos were identified, with Caatinga as the predominant vegetation type.

2. 3 Arbuscular mycorrhizal fungi in the recovery of degraded areas

The first reports on mycorrhiza appeared around 1885 by the German botanist Albert Bernard Frank (BAGYARAJ, 1991). According to their morphology and anatomy, mycorrhizae can be classified into two groups: ectomycorrhizae, in which the host root is covered externally by a thick mantle of hyphae, and the penetration of the internal mycelium into the root cortex is always intercellular, forming the so-called Hartig's net. And endomycorrhizae, which are characterized by the absence of a mantle of hyphae around the roots, but with intercellular and intracellular penetration of the inner mycelium into the cortex cells, forming specific structural units known as arbuscules and/or vesicles. They are of particular interest in the tropics where the vast majority of soils have low fertility and the formation of mycorrhiza is important for plant growth and survival, as well as for plant succession and the recovery of degraded areas (JANOS, 1996).

Arbuscular mycorrhizal fungi (AMF), which belong to the phylum Glomeromycota and class Glomeromycetes, are obligate symbionts that establish a mutualistic relationship with the roots of plants in the Gymnosperm, Angiosperm and some bryophyte and pteridophyte groups (SOUZA et al., 2010). Currently, around 19 genera have been described, with a total of 220 recognized species (OEHL et al., 2008; GOTO et al., 2010).

AMF represent one of the main components of the soil microbiota in most agroecosystems (SCHNEIDR et al., 2011). They provide numerous benefits to their hosts, such as increased tolerance to drought and phytopathogens, and are important in the absorption of nutrients such as phosphorus (P), nitrogen (N), copper (Cu) and zinc (Z). Their extraradicular hyphae play a fundamental role in the formation of aggregates and soil structure through the production of glomalin, a glycoprotein which, among other functions, contributes more effectively to the cementation of soil particles, thus increasing the stability of their aggregates (JOHNSON et al., 2010; KABIR, 2005).

The conservation and rehabilitation of degraded lands are fundamental parts of sustainable agricultural development and ecosystem improvement. In order to recover degraded areas, revegetation is considered a very important practice (AL-KARAKI, 2011). However, in semi-arid regions, long periods of drought make this practice difficult, leaving the soil bare and exposed for longer to the action of climatic agents, reducing its productive potential and often causing irreversible damage to the soil (TREVISAN et al., 2002; MENEZES et al., 2005). Therefore, the need to improve the environmental quality of these areas has boosted research into the association between plant development and microbial activity, an important factor in the recovery of degraded soils.

Arbuscular mycorrhizal fungi are found naturally in these areas (MASCHIO et al., 1992), where they are considered valuable tools for rehabilitation processes, as the symbiosis can confer numerous benefits on the host, exerting a great influence on plant growth and adaptation, including improved mineral nutrition as well as tolerance to diseases and abiotic stresses such as drought, salinity and high temperatures (AL-KARAKI, 2000).

Studying degraded semi-arid savannas in Africa, Veenendaal (1991) verified plant adaptation strategies and found that the presence of symbiosis with AMF was fundamental for the establishment of grass species, which favored the recovery of these areas.

Research carried out in the most diverse degraded situations (agricultural soils, deforested areas, mined areas, soils contaminated by metals and chemicals) has proven the importance and potential of mycorrhizae as supporting agents in the recovery of these degraded areas (BI et al., 2005; ZUQUETTE et al., 2013).

2. 4. Soil protein related to glomalin (PSRG)

PSRG is a hydrophobic, thermostable and recalcitrant substance produced by the decomposition of AMF hyphae (SOUSA et al., 2012) and can also be found in their spores and adsorbed on the roots of plants colonized by them (RILLIG & MUMMEY, 2006). The meaning of the word glomalin refers to the taxonomic order Glomales to which the AMF belonged at the time of their isolation and biochemical characterization.

PSRG, at first classified as a glycoprotein, is made up of 60% carbohydrates which are joined to the protein portion by N-type glycosidic bonds, with aliphatic and aromatic amino acids in the peptide chain. It has also been shown that iron is strongly bound to this molecule (0.04-8.8%), which is certainly the chromophore that transmits the brownish-red color observed in the extracts (RILLIG et al., 2001; NICHOLS, 2003).

This glycoprotein can contain from 3 to 5% N and a significant C content of around 37%, making it an important constituent in the soil's carbon stock (LOVELOCK et al., 2004). According to Wright & Anderson (2000) there is a strong relationship between the stability of soil aggregates and the concentration of PSRG, thus improving soil aeration, drainage and microbial activity, consequently promoting an improvement in the productivity of ecosystems.

Generally speaking, there are two types of PSRG: easily extractable (PSRGFE), which represents the fraction that has recently been deposited and has not yet undergone biochemical changes in the soil; and total PSRG (PSRGT), which is strongly adhered to the clays and requires up to seven sequential extractions for its maximum removal in some soils (WRIGHT et al., 1996; WRIGHT & UPADHYAYA, 1998).

According to Wright et al. (1996) glomalin can be extracted from soil or hyphae using 20 mM sodium citrate pH 7.0 (for PSRGFE) or 50 mM pH 8.0 (for PSRGT), by autoclaving at 121 °C and 1 atm pressure. Citrate acts as an iron chelator in conjunction with glomalin, removing the molecule adsorbed to the clay particles, while sodium breaks the iron and aluminium oxide bonds between the clay minerals and glomalin, helping to solubilize this molecule (WRIGHT et al., 2006). It was Wright and Upadhyaya (1998) who proposed the current protocol for extracting and quantifying glomalin fractions in soil.

PSRG is usually quantified using routine biochemical methods for protein dosage, such as the Bradford method. It can also be done using immunological techniques, such as the ELISA (Enzyme Linked Immuno Absorbent Assay) method, which uses a monoclonal antibody MAb32b11 produced from *Glomus intraradices* spore macerate injected into mice, which recognizes the glomalin epitope only in AMF and does not react with other soil fungi (WRIGHT et al., 1996). Reports have shown that polyphenolic compounds, such as soil tannins and humic acids, can be extracted together with glomalin and interfere with Bradford quantification (SCHINDLER et al., 2007). Bearing this in mind, in order to adequately differentiate the specific protein (glomalin) from the mixture of proteins resulting from extraction with sodium citrate, Rillig (2004) proposed using the term soil protein related to glomalin (PSRG) to describe the extracted mixture, leaving the term "glomalin" reserved only for the purified protein.

2. 5. 1 Other glomalin-related soil protein contributions

According to Purin and Rillig (2007) glomalin is probably not restricted to improving the stability of soil aggregates. Cornejo et al. (2008) show that glomalin can be a highly efficient tool for sequestering heavy metals, reducing the availability of these elements for organisms and plants in contaminated soils. Purin & Rillig (2007) also showed that the production of this glycoprotein by AMF increases in soils polluted by trace elements, probably as a mechanism to improve the environment in which the symbiotic fungus develops. Bedini et al. (2009) showed that the amounts of Cu, Ni, Pb and Co bound to GRSP were 2, 3, 0.83 and 0.24 % respectively of the total heavy metal content in contaminated soil, thus reducing the availability of trace elements and, consequently, the plant's stress from these metals.

Another contribution of glomalin in the soil would be in the stock of C and N, where it is estimated that between 28 and 45% of the glomalin molecule is carbon and it is presumed that 0.9 to 7.3% is nitrogen, thus evidencing a source of these elements for the soil microbiota (RILLIG et al., 2003; LOVELOCK et al., 2004; NICHOLS & WRIGHT, 2006). Lovelock et al. (2004) found that the C and N present in glomalin corresponded to approximately 3.2 and 5% of the total C and N pool in two soils at the

La Selva biological station in Costa Rica. These differences in the contribution of glomalin vary according to the type of soil and the type of land use system (SOUSA et al., 2012).

3. MATERIAL AND METHODS

3. 1. location of the study area

The study was carried out in the experimental area of the project "Studies of Degradation/Desertification Processes and their Relationships with Land Use in Production Systems in the Semi-Arid Region of Ceará: The Case of the Micro-Region of Sobral - Cearà". The area is located on the Formigueiro farm in the municipality of Irauçuba in the north-central hinterland of the state of Ceará, part of the Sobral micro-region, at coordinates 03° 46' 50" S and 39° 49' 03" W (Figure 1). The municipality covers an area of 1,451 km^2 , with an altitude of 152.5 meters. The region's climate is semi-arid, with an average annual temperature and rainfall of 26.3 °C and 530 mm respectively, concentrated basically in three months of the year.

Figure 1 - Map of the soil collection points in the three areas located in Irauçuba-CE.

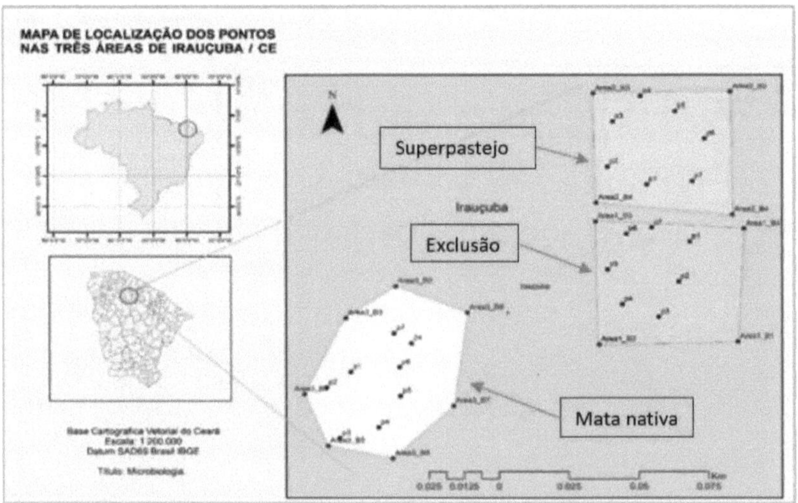

Source: Modified IBGE.

3. 2. Soil collection

Soil samples were taken from three areas of 0.25 hectares each (Figure 2) and (Figure 3), characterized by different levels of degradation and recovery: Overgrazed area, with little or no natural vegetation and severe laminar erosion; Exclusion area, which remains in the process of natural regeneration, free from domestic animals, fenced off with nine strands of barbed wire and stakes every meter; Native forest area, characterized by shrubby-arborea caatinga vegetation. Seven deformed soil samples were collected from each area, chosen randomly by zigzag walking and georeferenced to a depth of 0-10 cm.

Figure 2 - Aerial view of the areas where the soil samples were collected.

Source: google maps.

Figure 3: Photo of the experimental area - exclusion and its surroundings (overgrazing)

Source: author.

3. 3. Physical analysis of the soil

The physical analyses carried out were: aggregate stability using the wet method, the principle of which is to measure the quantity and size distribution of aggregates that are stable in water, relating them to those that do not disintegrate by sieving; particle size analysis, determined using the pipette method (GEE & BAUDER, 1986), dispersing 20 g of TFSA with 1 mol L sodium hydroxide^{-1} . The sands were separated on 0.053 mm mesh sieves and fractionated according to the United States Department of Agriculture (USDA) reference particle size classification. The silt and clay were separated by sedimentation according to Stokes' law.

Table 1. Soil physical characteristics: granulometry and aggregate stability in three areas: overgrazed, excluded and native forest in the municipality of Irauçuba-CE. 2015.

Areas/Treatments	Sand (%)	Silt (%)	Clay (%)	AS (%)	Classification
Overgrazing	77,87	10,01	12,12	74,43	Sandy loam
Exclusion	79,00	10,70	10,30	81,91	Sandy loam
Native Forest	68,54	15,14	16,32	85,43	Sandy loam

EA: aggregate stability.

3. 4. Soil chemical analysis

The chemical analysis of the soil samples was carried out at the Soil-Water-Plant Laboratory of the Soil Sciences Department of the Federal University of Ceará (UFC), where the following were evaluated: pH (in H_2O) by measuring with a combined electrode immersed in soil suspension; potential acidity (H^+ Al^{3+}) which was obtained by extraction with buffered calcium acetate solution pH 7.0; exchangeable aluminum (Al^{+3}) was extracted with KCl solution and determined volumetrically with diluted NaOH solution; exchangeable bases (Ca^{+2} , Mg^{+2}) determined by the KCl method; phosphorus (P- available) using an extractant solution (HCl 0.05 N and H_2SO_4

0.025 N) as well as an acidic solution of diluted ammonium molybdate and ascorbic acid powder as a reducer, and the reading was taken by photocolorimetry; organic matter (OM) and organic and total carbon (OC and TOC) were determined by oxidizing the organic matter with potassium dichromate (K2Cr2 0.167 mol/l^{-1}) in a sulphuric medium (H_2SO_4). After oxidation, the excess dichromate was titrated with ammoniacal ferrous sulphate solution $Fe(NH_4)_2(SO_4).6H_2O$ 0.5 mol/l^{-1} in the presence of the indicator diphenylamine and the organic matter content was then estimated. All these methodologies are described in EMBRAPA's Manual of Soil and Plant Analysis Methods (1997).

Table 2. Average values of soil chemical attributes at depths of 0-10 cm in areas of overgrazing, exclusion and native forest in the municipality of Irauçuba-CE. 2015.

Attributes	Sample areas		
Chemicals	Overgrazing	Exclusion	Native Forest
0-10 cm			
pH (H_2O)	4,96	5,00	5,69
Ca^{2+} (cmolc kg)$^{-1}$	2,37	2,41	6,60
Mg^{2+} (cmolc kg)$^{-1}$	2,07	1,79	3,37
Al^{3+} (cmolc kg)$^{-1}$	0,21	0,37	0,10
H^{+} Al^{3+} (cmolc kg)$^{-1}$	2,73	4,07	4,77
P (mg kg)$^{-1}$	66,18	85,60	18,41
CO (g kg)$^{-1}$	4,52	7,39	9,61
N (g kg)$^{-1}$	0,64	0,77	1,09
M.O. (g kg)$^{-1}$	7,78	12,70	16,53

3. 5. Spore density in the soil

100 g of soil were taken from the seven samples in each area to extract the AMF spores using the wet sieving technique (GERDERMANN & NICOLSON, 1963). A set of overlapping sieves (with mesh sizes varying between 850, 250, 100 and 45 μm) were used to extract the total glomerospores. The spores were then quantified on a channeled plate using an optical stereoscopic microscope (40x).

3. 6. Extraction and quantification of soil protein related to glomalin (PSRG).

PSRG was extracted using the method proposed by Wright & Upadhyaya (1998), in which the easily extractable PSRG fractions - $PSRG_{FE}$, obtained from sequential extraction in an autoclave, were extracted by adding 1 g of soil (Figure 4) in 8 mL of 20 mM sodium citrate buffer solution (pH 7,0), at a temperature of 121° C for 30 min, and the total PSRG fraction - $PSRG_T$, obtained by sequential extraction from 1 g of soil added to 8 mL of 50 mM sodium citrate buffer solution (pH 8.0) at 121° C for 60 min. After autoclaving, both fractions were centrifuged at 10000 rpm for 5 min, when the supernatant was removed for subsequent protein quantification (Figure 5). Glomalin was quantified by the Bradford method (1976), using bovine serum albumin as the protein standard. This method used the dye Comassié Brilliant Blue G-250 (CBB or Bradford Reagent), which has the ability to form complexes with the proteins in the solution, causing a change in absorbance, which is equivalent to the amount of protein present (BRADFORD, 1976).

Figure 4: Tubes containing 1 g of soil before PSRG extraction.

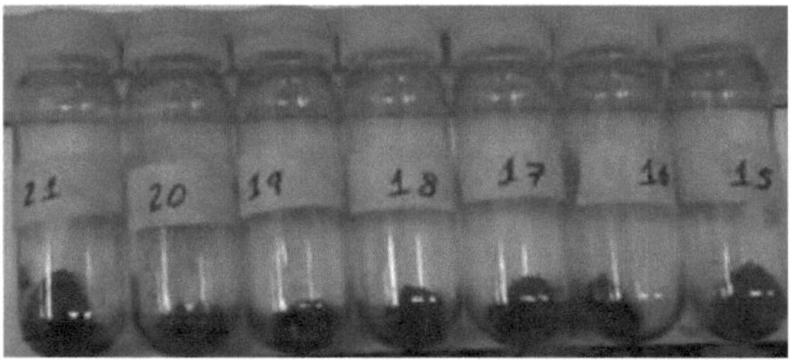

Source: author.

Figure 5 - Vials containing PSRGFE in the supernatant immediately after autoclaving and centrifugation.

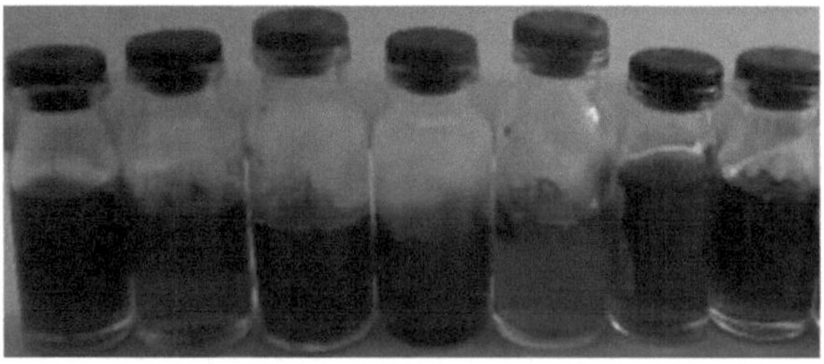

Source: author.

3. 7. Statistical analysis

A descriptive statistical analysis of the data was carried out by comparing means using the Assistat program. Data was also tested for normality at 5% probability using the Shapiro-Wilk (W) test. The data was then analyzed using multivariate statistical tools by means of canonical correlation analysis, adopting two groups: Group one (I), variables related to physical and biological chemical attributes (pH, Ca^2 +, Mg^2 +, Al^3 +, P, CO, N, EA, DE), and group (II), variables made up of the two glomalin-related soil protein fractions (PSRGFE and PSRGT). Another multivariate statistical tool used was hierarchical cluster analysis.

Statistica® software version 7.0 was used for all multivariate analyses.

4. RESULTS AND DISCUSSION

4. 1 PSRG fractions in the three areas: overgrazing, exclusion and native forest.

The highest average PSRGFE and PSRGT concentrations were found in the native forest area (Table 2). In the areas of exclusive overgrazing, there was no statistically significant difference between the two fractions of PSRG.

TABLE 3: PSRGFE and PSRGT content (mg g^{-1} soil) in soil samples from degraded areas (Overgrazed), under management (Exclusion) and native forest in the municipality of Irauçuba-CE. 2015.

AreasZTreatments	PSRGFE	PSRGT
Overgrazing	0,22b	0,18b
Exclusion	0,54b	0,44b
Native forest	0,93a	0,83a

Means followed by the same letter in the column do not differ by the Scott Knott test (p <0.01).

Several authors suggest that this protein is deposited in the soil mainly through the decomposition of AMF hyphae and spores (DRIVER; HOLBEN; RILLIG, 2005). Rillig et al. (2003) also proved that PSRG concentrations are related to organic matter in the soil, as the higher the organic matter content, the higher the PSRG values found. In this study, the native forest was the area with the highest mean values for spore density (SD) (Table 4) and organic matter (OM) content (Table 2), so the fact that the native forest had the highest PSRG concentrations could possibly be attributed to the higher SD and OM content observed.

The other areas (overgrazed and excluded) did not show significant differences in terms of the content of the two fractions of PSRG (Table 3). However, it is important to note that the overgrazed area showed the lowest average quantitative concentration values of the PSRG fractions, despite the lack of statistical significance in relation to the exclusion area. Some authors state that human activities such as the removal of natural vegetation and the inappropriate use of the soil, combined with climatic factors such as light intensity and temperature, impair the development of hyphae and spores which are responsible for the production of PSRG (MERGULHÂO, 2006;

CARDOZO, 2011). The overgrazed area studied has little or no vegetation and a consequent process of degradation, with the presence of laminar erosion and inserted in a context of adverse climatic conditions, which may have hindered the production of external mycelium in the soil, the main factor responsible for the synthesis of PSRG.

In the exclusion area, higher average quantitative values were observed in the contents of the two PSRG fractions compared to the overgrazed area, despite the lack of statistical significance (Table 3). It is important to note that it is possible that the exclusion, being an area in the process of natural regeneration with shrub-type vegetation and having higher mean values for spore density and organic matter content than the overgrazed area, may have contributed to stimulating the occurrence of AMF and, consequently, increasing the deposition of this protein in the soil (WRIGHT & UPADHYAYA, 1996).

4. 2 Aggregate stability and spore density in the three areas: overgrazing, exclusion and native forest.

When comparing the three areas, the exclusion and native forest did not show significant differences in terms of average aggregate stability values, with these two areas obtaining the best percentages of stable aggregates in the soil (Table 4).

TABLE 4: Aggregate stability (AS) and spore density (SD) in soil samples from degraded areas (Overgrazed), under management (Exclusion) and native forest in the municipality of Irauçuba- Ce. 2015.

Areas/Treatments	AS (%)	DE (100mL/soil)
Overgrazing	74b	4269b
Exclusion	82a	5031b
Native forest	85a	7833a

Means followed by the same letter in the column do not differ by the Scott Knott test ($p<0.01$).

According to some authors, the stability of soil aggregates is very sensitive to changes in ecosystems and can be directly altered by the soil management systems adopted (BARTO et al., 2010; SOUSA, 2009). This fact can be observed in this study in the overgrazed area, where the lowest average aggregate stability value was obtained (Table 4), and this result can be attributed to the degree of degradation in which the area is located. Bird et al. (2007), comparing aggregate stability in a degraded area with a non-degraded area in the semi-arid region of New Mexico, found that aggregate stability behaved according to the degree of degradation, being more stable in areas with greater vegetation cover, since organic matter has a significant influence on soil aggregation.

With regard to the spore density of the overgrazed and excluded areas, there was no statistically significant difference, with the lowest average spore values (Table 3) compared to the native forest. In the case of these two areas, the overgrazed area remains in degraded conditions, as it has little or no natural vegetation and severe laminar erosion, while the exclusion area, which was conceived as a model for recovering these areas, has seen very little in the way of significant variations in the typical vegetation over these years of fallow (less than 20 years) (VALONE & SAUTER, 2005; PEI et al., 2007; ZHAO et al., 2007). According to McLellan et al. (1995), the propagules of arbuscular mycorrhizal fungi have a close relationship with the processes related to land degradation that occur in various regions of the world, being severely affected when the soil-plant interaction is affected by environmental imbalance. However, Sylvia & Williams (1992) point out that a more stressful environmental condition leads AMF to produce a high number of propagules in order to survive, a fact that was not observed in this study, since the native forest had the highest number of spores. This was to be expected as this area has greater plant diversity, in which plants form associations with arbuscular mycorrhizae, promoting colonization and sporulation, as this is dependent on root colonization (FRANKE & MORTON, 1994).

4. 3 Multivariate analysis: canonical correlations of chemical, physical and biological attributes with PSRG fractions.

Table 5 shows the correlations and canonical pairs between physical, chemical and biological attributes related to the concentrations of the two fractions of PSRG ($PSRG_{FE}$ and total-PSRGT), in three different areas (overgrazed, excluded and native forest) in Irauçuba-CE. It can be seen that for the first pair, the canonical correlations were highly significant (1% probability) using the chi-squared test (0.97). Therefore, set (I) explains set (II) in the first canonical pair and is therefore of interest to the study. Among the variables related to the concentration of PSRG, the attributes that contributed most to explaining the concentration of PSRG in the soil were Ca^{2+} (R=0.86), N (R=0.84), CO (R=0.84), Mg^{2+} (R=0.71), pH (R=0.70) and EA (R=0.61), respectively. In other words, as the values of set (I) increased, so did the concentration of PSRG-FE (R=0.94) and PSRG-T (R=0.99), in the following order in the soil; on the other hand, there was little or no effect (R<0.6) for the attributes Al^{3+} (R=0.45), P (R=0.08) and DE (R=0.18). For the second canonical pair, no significant differences were found between the attributes of (CI) and PSRG concentration (CII) in the soils of Irauçuba-CE.

Table 5. Canonical correlations and canonical pairs between the characteristics of the CI (pH, Ca, Mg, Al, P, CO, N, EA and DE) and CII (PSRGFE and PSRGT) sets for the 21 study points.

Canonical Pairs

Sets	Attributes	Canonical Pairs	
		1	2
		Canonical correlations	
	pH	-0,701274	0,078037
	Ca^{2+}	-0,863931	0,094888
	Mg^{2+}	-0,714206	0,304614
	Al^{3+}	0,456270	0,028860
CI	P	-0,081694	-0,137169
	CO	-0,841806	0,062830
	N	-0,847321	0,053432
	EA	-0,617521	-0,081882
	DE	0,180468	0,118810
IIC	PSRGFE	-0,942104	-0,335321
	PSRGT	-0,998485	0,055027
	R-Canonical	0,9729795**	$0,8345262^{ns}$

Chi-square	55,67089	16,09411
GL	20	9

GL: Degrees of freedom; **, *, ns, significant at 1%, 5% and not significant respectively. (1) Canonical correlations ≥ 0.6 were considered significant for interpretation purposes.

Calcium (Ca^2 +) was the variable in group I that had the highest correlation with the variables in group II in three areas (Table 4), thus demonstrating the importance of calcium in the production of PSRG in this study. Andrade et al. (1995), studying coffee cultivation on purple latosol in the state of Paranà, observed that mycorrhizal colonization increased with the addition of limestone. In the present study, Ca^{2+} showed a high correlation (0.86) with the concentration of both types of PSRG, but its highest correlation was with PSRGT. WU et al. (2014) state that Ca^{2+} may participate in the establishment of arbuscular mycorrhizae (AM), promoting the growth of hyphae and spores which, with the consequent decomposition of these fungal structures, will lead to the formation of PSRG, a fact which may explain the strong correlation between calcium and the protein produced by glomalean fungi. However, these results are controversial and many other authors have shown a negative relationship or lack of correlation between this protein and Ca^2 + levels (LOVELOCK et al., 2004; MERGULHÂO, 2006).

In decreasing order of correlation, N and organic C were the variables which, after Ca^2 +, showed the highest correlations (0.84) with the two types of PSRG (Table 4). However, of the two PSRG fractions, PSRGT was the one with the highest correlation with the organic N and C variables, which can probably be explained by the fact that this fraction is more stable and, consequently, has a reduced decomposition rate, which brings higher concentrations of these two elements into the soil (RILLING et al., 2001; STEINBERG & RILLING, 2003). The positive correlation between the organic N and C variables and the two fractions of PSRG can be attributed to the fact that this glycoprotein has considerable amounts of carbon and nitrogen in its chemical composition, and can contain estimated values of between 28 and 45% C and 0.9 to 7.3% N, thus constituting an appreciable reservoir of C and N in soils (RILLIG et al., 2003;

LOVELOCK et al., 2004; NICHOLS & WRIGHT, 2006). Organic C is one of the most consistent indicators of the concentration of PSRG in ecosystems (SOUSA et al., 2011). Positive correlations between organic C and PSRG have been demonstrated (BIRD et al., 2002; FRANZLUEBBERS, 2000). Lovelock et al. (2004), working with two Costa Rican soils, found that the C and N contents in the PSRG corresponded to approximately 3.2 and 5% of the total C and N pool of these soils. Similar results were found by Rillig et al. (2001).

Mg^{2+} also showed a correlation with PSRG (Table 4) and may have the same explanation as the correlation with Ca^{2+} , since these two elements behave very similarly in the soil due to their chemical similarity, since they are both divalent cations and their ionic sizes are also close to each other. It is probable that Mg^{2} + also participates in the establishment of AMF and these are responsible for the production of PSRG (HODGES et al., 1992). In the present study, the values found for Mg^{2+} were in line with those obtained by Silva et al. (2012) and also correlated positively with PSRG.

pH was another variable in set (I) that showed a positive correlation (0.70) with the two fractions of PSRG in set (II) (Table 4). Fungi tend to predominate in acidic soils and, as PSRG is produced by AMF, it is to be expected that there will be greater production of this protein in acidic soils (HADDAD & SARKAR, 2003). It is possible that in acidic pH conditions there is greater preservation of PSRG, since this condition is not favorable to the presence and activity of microorganisms that decompose this molecule, such as actinobacteria (LIMA et al., 2013). However, some authors have observed a negative correlation between glomalin levels and soil pH (RILLING et al., 2003; HADDAD & SARKAR, 2003; MERGULHÂO, 2006).

Several studies have shown that the concentration of PSRG has a high correlation with soil aggregate stability, consequently promoting improvements in soil structure and edaphic quality and contributing to the reduction of erosion risks (WRIGHT; UPADHYAYA, 1998; WRIGHT; ANDERSON, 2000; WRIGHT et al., 2007). In this study, a positive correlation (0.61) was observed between the two fractions of PSRG and aggregate stability (Table 4). It is believed that the relationship between PSRG and aggregate stability is attributed to the recalcitrance and hydrophorbicity of this protein, which acts as an organic binder, helping to fix particulate material in the soil, as well as favoring the formation and stability of aggregates (PURIN, 2005; WRIGHT et al., 1996). Nobre et al. (2015), studying soil organic carbon levels and PSRG in different phytophysiognomies in the Caatinga biome, in the Chapada do Araripe (CE), also found a positive correlation between the protein produced by AMF and the stability of soil aggregates.

On the other hand, the variables in set I, Al (R=0.45), DE (R=0.18) and P (R=0.08), showed little correlation with the variables in set II ($PSRG_{FE}$ and $PSRG_{Tz4}$), since the respective variables Al, DE and P obtained canonical correlations with values below 0.6 in modulus, which according to Santos (2010), variables with values <0.6 are the ones that contribute least to explaining the correlations between the variables in the sets formed.

4. 4. Hierarchical cluster analysis of all the variables analyzed in the three study areas.

The dendogram shows the formation of three affinity groups (Figure 6). It is important to note that the cluster analysis procedure looks for the characteristics of the same group that make them similar, while differing from the behavior of the other groups (VALLADARES et al., 2008). The first group was formed by the native forest area, the second by the overgrazed area and the third by the exclusion area. The distinction between the native forest area and the overgrazed and excluded areas was thus evident, as was the similarity between the latter two (overgrazed and excluded), proving that more time is needed for significant differences to occur between these areas in terms of recovery.

Figure 6. Dissimilarity between groups established by Euclidean distance based on chemical attributes: (pH, Ca^2 +, Mg^2 +, Al^3 +, P, CO, N); physical: (EA) aggregate stability and biological: (DE) spore density and soil protein related to glomalin (PSRG) (easily extractable-$PSRG_{FE}$ and total-$PSRG_T$), in soils from three areas: degraded (Overgrazing-SP), under management (Exclusion-EX) and native forest-MN, in the municipality of Irauçuba-CE. 2015.

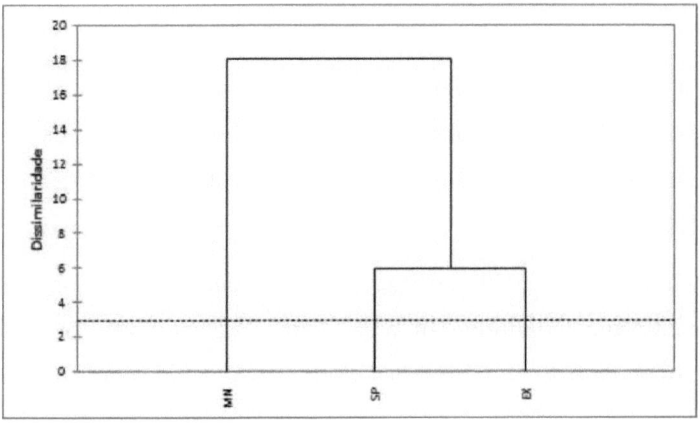

Source: author.

5. CONCLUSIONS

The content of the soil protein fractions related to glomalin (easily extractable and total) in the native forest were higher than those found in the exclusion and overgrazed areas, proving that the intense degradation of these areas contributed to a decrease in the concentration of PSRG.

The content of PSRG, associated with other edaphic attributes, contributes to discriminating the quality and monitoring of areas with different levels of soil degradation in Irauçuba-CE.

The hierarchical cluster analysis based on all the parameters assessed in the soil showed greater similarity between the overgrazed and excluded areas and greater dissimilarity between them and the native forest, indicating that more time is needed for significant differences to occur between these areas in terms of their recovery.

6. REFERENCES

ABNT. BRAZILIAN ASSOCIATION OF TECHNICAL STANDARDS. **NBR 10703**: Soil degradation. Sao Paulo, 1989.

ACCIOLY, L. J. O. Soil Degradation and Desertification in Northeast Brazil - **Portal Dia de campo**. Available at: <http://www.diadecampo.com.br> Accessed in: July 2014.

AL-KARAKI, G. N. Mycorrhizal fungi role in reducing the environmental impacts of climate change in arid regions. In: Solh M, Saxena M (ed) food security and climate change in dry areas: **Proceedings of the International Conference on 1 to 4 February 2010**, Amman, Jordan International Center for Agricultural Research in Dry Areas, Aleppo, Syria, p. 304 -313. 2011.

AL-KARAKI, G. N. Growth of mycorrhizal tomato and mineral acquisition under salt stress. **Mycorrhiza**, v. 10, p. 51-54, 2000.

ANDRADE, D.S. et al. Microbial activity as a function of liming in a soil cultivated with coffee trees. **Revista Brasileira de Ciência do Solo**, Campinas, v. 19, p. 191-196, 1995.

ARAÙJO, E. A.; KER, C. J.; NEVES, J. C. L.; LANI, J. L. Soil quality: concepts, indicators and evaluation. **Revista Brasileira de Tecnologia Aplicada nas Ciências Agràrias**, Guarapuava-PR, v. 5, n. 1, p. 187-206, 2012.

ARAÙJO, G. H. de S.; ALMEIDA, J. R. de; GUERRA, A. J. T. **Gestao Ambiental de Areas Degradadas.** 3ª Ed. Rio de Janeiro: Bertrand Brasil, 2008. 320p.

BAGYARAJ, D. J. Ecology of vesicular-arbuscular mycorrhizae. In: ARORA, D.K.; RAI, B.; MUKERJI, K.G.; KNUDSEN, G.R. (Eds).

Handbooke of applied mycology: soil and plant. New York. Marcel Dekker. v. 1, p. 4-34, 1991.

BARTO, E. K.; ALT, F.; OELMANN, Y.; WILCKE, W.; RILLIG, M. C. Contributions of biotic and abiotic factors to soil aggregation across a land use gradient. **Soil Biology and Biochemistry,** v. 42, n. 12, p. 2316-2324, 2010.

BATISTA, Q. R.; FREITAS, M. S. M.; MARTINS, M. A.; SILVA, C. F. Bioquality of an area degraded by clay extraction, revegetated with *Eucalyptus* spp. and sabià. **Revista Caatinga**, Mossoró, v. 22, n. 1, p. 146154, jan./mar. 2009.

BEDINI, S.; PELLEGRINO, E.; AVIO, L.; PELLEGRINI, S.; BAZZOFFI, P.; ARGESE, E.; GIOVANNETTI, M. Changes in soil aggregation and glomalin-related soil protein content as affected by the arbuscular mycorrhizal fungal species *Glomus mosseae* and *Glomus intraradices*. **Soil Biology Biochemistry**. v. 41, n. 7, p. 1491-1496, 2009.

BI, Y. L; WU, F. Y; WU, Y. K. Application of arbuscular mycorrhizal in: ecological restoration of areas affected by mining in China. **Acta Ecologica Sinica.** v. 25, p. 2068-2073, 2005.

BLUNDEN, J.; ARNDT, D. S. **State of the Climate 2011**. Bulletin of the American Meteorological Society, 2012. v. 93, n. 7, p. 1-264.

BRADFORD, M. M. A rapid and sensitive method for the quantification of microgram quantities of protein utilizing the principle of protein-dye binding. **Analytical Biochemistry**, New York, v. 72, n. 1-2, p. 248-254, May. 1976.

BRAZIL. Ministry of Agriculture. National Agricultural Research Department. Pedological Division. **Exploratory Survey**: Soil Reconnaissance of the State of Ceará. Recife, 1973. v. 1. (DNPA. Boletim Tècnico, 28, SUDENE. Série Pedologia, 16)

BRAZIL. Secretariat for Regional Development Policies. **New delimitation of the Brazilian semi-arid** region. Ministry of National Integration. Brasilia, 2005.

BRAZIL. Ministry of the Environment. Secretariat for Water Resources. **National action program to combat desertification and mitigate the effects of drought**:

Brasilia, DF, p. 213, 2004. PAN-BRASIL.

CARDOZO JR, F. M. **Dynamics and biodiversity of arbuscular mycorrhizal fungi in different areas under the influence of the desertification process in Gilbués-Brazil**. 2011. Federal University of Piaui. Piaui- Gilbués. Dissertation (Master's Degree in Agronomy), 2011.

CENTORION, J. F.; CARDOSO, J. P.; NATALE, W. Effect of management methods on some physical and chemical properties of a Red Latosol in different agroecosystems. **Brazilian Journal of Agricultural and Environmental Engineering**. Campina Grande, PB, v. 5. p. 254258, 2001.

CORNEJO, P.; MEIER, S.; BORIE, G.; RIILING, M. C.; BORIE, F. Glomalin-related soil protein in a Mediterranean ecosystem affected by a copper smelter and its contribution to Cu and Zn sequestration. **Science of the Total Environment**, v. 406. p. 154-160, 2008.

DORAN, J. W.; COLEMAN, D. C.; BEZDICEK, D. F.; STEWART, B. A.

Defining soil quality for a sustainable environment, **Soil Science Society of America**, Inc. Issue 35, Madison, WI, USA, p. 3-21, 1994.

DRIVER, J. D.; HOLBEN, W. E.; RILLIG, M. C. Characterization of glomalin as a hyphal wall component of arbuscular mycorrhizal fungi. **Soil Biology & Biochemistry**, Oxford, v. 37, n. 1, p. 101-106, 2005.

DRUMOND, M. A. et al. Strategies for the sustainable use of caatinga biodiversity. In: **Workshop to evaluate and identify priority actions for the conservation, sustainable use and sharing of benefits from the biodiversity of the caatinga biome.**

Petrolina, 2000. 23 p.

DRUMOND, M.A.; KIILL, L. H. P.; LIMA. P. C. F.; OLIVEIRA, M. C.; OLIVEIRA, V. R.; ALBUQUERQUE, S. G.; NASCIMENTO, C. E. S.; CAVALCANTI, J.

Estratégias para o uso sustentàvel da biodiversidade da caatinga. In: SILVA, J. M. C. da; TABARELLI, M.; FONSECA, M. T.;

LINS, L. V. (Org.) **Biodiversity of the caatinga: priority areas and actions for conservation. Brasilia**, DF: Ministry of the Environment: Federal University of Pernambuco, 2004. 329-340 p.

EMBRAPA - BRAZILIAN RESEARCH COMPANY

AGROPECURIA. **Manual of soil analysis methods**. Rio de Janeiro: EMBRAPA, 1997. 212 p.

BRAZILIAN AGRICULTURAL RESEARCH Corporation - EMBRAPA. National Soil Research Center. **Brazilian Soil Classification System. 2ª Ed**. Rio de Janeiro. 2006. 306p.

FAO. Food and Agriculture Organization of the united Nations. **Symposium on land degradation and poverty**. Rome, italy, FAO, international Fertilizer industry Association-iFA, 2000, p. 3. At: <http://www.fao.org/docrep/X5317E/x5317e00.htm>. Accessed on: July 23, 2014.

FRANKE, M.; MORTON, J. B. Ontogenetic comparisons of arbuscular mycorrhizal fungi scutellospora heterogama and scutellospora pellucida: revision of taxonomic character concepts, species descriptions, and phylogenetic hypotheses. **Canadian Journal of Botany**. v. 72, p. 122-134, 1994.

GAuR, A.; VARMA, A. Research methods in arbuscular Mycorrhizal fungi. in: VARMA, A.; OELMÜLLER, R. (eds). **Soil biology**: **Advanced techniques in soil microbiology**. springer-Verlag Berlin Hheidelberg, p.376-396, 2007.

GERDEMANN, J. W.; NiCOLsON, T. H. spores of mycorrhizal endogon species extracted from soil by wet sieving and decanting. **Transactions of the British Mycological Society**, London, v. 46, p. 235-244, 1963.

GONÇALVEs, J. L. de M.; NOGuEiRA JuNiOR, L. R.; DuCATi, F. Recovering degraded soils. In: KAGEYAMA, Paulo Yoshio; OLiVEiRA, Renata Evangelista de; MORAEs, Luiz Fernando Duarte de; ENGEL, Vera Lex; GANDARA, Flàvio Bertin. **Ecological restoration of natural ecosystems**. Botucatu, SP. FEPAF, Cap.06.p.-113-163, 2003.

GOTO, B.T.; SILVA, G. A.; YANO-MELO, A. M.; MAIA, L. C. Checklist of the arbuscular mycorrhizal fungi (*Glomeromycota*) in the Brazilian semiarid. **Mycotaxon**, v. 113, p. 251-254, 2010.

GREGORICH, E. G.; CARTER, M. R.; ANGERS, D. A.; MONREAL, C. M.; ELLERT, B. H. Towards a minimum data set to asset soil organic matter quality in agricultural soils. **Canadian Journal of Soil Science**. Volume 74, Issue 4, Canada (1994), p. 367-385, 1994.

HADDAD, M. J.; SARKAR, D. Glomalin, a newly discovered component of soil organic matter: part II - relationship with soil properties.

Environmental Geosciences, v. 10, n. 3, p. 99-106, 2003.

HODGES, S.C., AND G. J. GASCHO. Comparison of plow layer and pegging zone soil test results in Georgia. **Proc. Am. Peanut Res. Educ. Soc**. v. 24, p. 31, 1992.

INSA. National Semi-Arid Institute. **Desertification and climate change in the Brazilian semi-arid** region. Campina Grande: INSA-PB, 2011. 209p.

JANOS, D. P. Mycorrhizas, succession and rehabilitation of deforested lands in the humid tropics. In: FRANKLAND, J. C.; GADD, G. M. (Ed.). **Fungi and environmental change**. Cambridge: Cambridge University Press, p. 1-18. (British Mycological Society Symposium, v. 20), 1996.

JOHNSON, N. C.; WILSON, G. W. T.; BOWKER, M. A.; WILSON, J. A.; MILLER, R. M. Resource limitation is a driver of local adaptation in mycorrhizal symbioses. **Proceedings of the National Academy of Sciences of the United States of America**. v. 107, p. 2093-2098, 2010.

KABIR, Z. Tillage or no-tillage: impact on mycorrhizae. **Canadian Journal of Plant Science**, Canada. v. 85, p. 23-29, 2005.

KARLEN, D. L.; MAUSBACH, M. J.; DORAN, J. W.; CLINE, R. G.;

HARRIS, R. F.; SCHUMAN, G. E. Soil quality: a concept, definition and framework for evaluation. **Soil Science Society America Journal**, v. 61, p. 4-10, 1997.

KASSAS, M. Desertification - a general-review. **Journal of Arid Environments.** v. 30, p. 115-128, 1995.

KENNEDY, B. K; STEFFEN, K. K; KAEBERLEIN, M. Ruminations on dietary restriction and aging. **Cell Molecules in Life Science.** v. 64, p. 1323-1328, 2007.

LIMA, F. S.; SOARES, A. C. F.; SOUSA, C. S. Occurrence and activity of arbuscular mycorrhizal fungi in *eucalyptus* (*Eucalyptus* sp.) plantations on the north coast of Bahia, Brazil. **Revista Arvore**, Viçosa-MG, v. 37, p. 245-255, 2013.

LOVELOCK, C. E.; WRIGHT, S. F.; CLARK, D. A.; RUESS, R. W. Soil stocks of glomalin produced by arbuscular mycorrhizal fungi across a tropical rain forest landscape. **Journal of Ecology**, v. 92, n. 2, p. 278-287, 2004.

LOVELOCK, C. E.; WRIGHT, S. F.; NICHOLS, K. A. Using glomalin as an indicator for arbuscular mycorrhizal hyphal growth: an example from a tropical rain forest soil. **Soil Biology & Biochemistry,** Oxford, v. 36, p. 1009-1012, 2004.

MAFRA, N. M. C. Erosion and Land Use Planning. In: GUERRA, Antonio José Teixeira; SILVA, Antonio Soares da; BOTELHO, Rosangela

Erosion and soil conservation: concepts, themes and applications. Rio de Janeiro: Bertrand Brasil, p. 301-322, 1999.

MAJER, J. D. Fauna studies in land reclamation technology: a review of the history and need for such studies. In: **Animals in primary succession: the role of fauna in reclaimed land**. J. D. Majer, coordinator, p 3-33.

London, Cambridge University Press. 1989.

MASCHIO, L.; GAIAD, S.; MONTOYA, L.; CURCIO, G. R.; RACHWALL, M. F. G.; CAMARGO, C. M. S.; BATTI, A. M. B. Microorganisms and self-sustainability of ecosystems in disturbed soils. In: NATIONAL SYMPOSIUM ON THE RECOVERY OF DEGRADED AREAS, Curitiba. **Proceedings**... Curitiba: UFPR/Fundaçao de Pesquisas Florestais do Paranà, p.440-445, 1992.

MATALLO JR. H.; Desertification in the world and in Brazil. In:SCHENKEL, C.S.; JÛNIOR, H.M. **Desertification-org**. UNESCO, 1999-2003. 82 p.

MCLELLAN, A.J., FITTER, A.H., LAW, R. On decaying roots, mycorrhizal colonization and the design of removal experiments. **Journal of Ecology**. v. 83, p. 225-230, 1995.

MEKURIA, W., VELDKAMP, E., HAILE, M., NYSSEN, J., MUYS, B., GEBREHIWOT, K. Effectiveness of exclosures to restore degraded soils as a result of overgrazing in Tigray. Ethiopia. **Journal of Arid Environments**. v. 69, p. 270-284. 2007.

MENEZES, R. S. C.; SAMPAIO, E.V.S.B. Simulation of phosphorus flows and balances in a family farm in the semi-arid region. Paraibano. In: SILVEIRA, L.M.; PETERSEN, P. & SABOURIN, E., eds. Family farming and agroecology in the semi-arid region:

Advances from Agreste da Paraiba. Rio de Janeiro, **AS-PTA**, p.249260, 2002.

MENEZES, R. C. S.; GARRIDO, M.S. & PEREZ, M. A. M. Fertility of soils in the semi-arid region. In: **Brazilian Congress of Soil Science**, 30. Recife, Lectures. Recife, UFRPE/SBCS, CD-ROM, 2005.

MERGULHÂO, A. C. E. S. **Ecological and molecular aspects of arbuscular mycorrhizal fungi**. Federal University of Pernambuco, Recife. 2006. 152 p. (Doctoral Thesis in Biological Sciences)

MESQUITA, R. C. M.; VALE, L. V.; LOPES, E. A. & OLIVEIRA, J. G. B. Management of native pastures for goats. In: I Simpósio Nacional de Caprinos e Ovinos Tropical, 1981, Sobral - CE. CNPC/EMBRAPA, 1981.

MOREIRA, P. R. **Manejo do solo e recomposição da vegetação com vista a recuperação de áreas degradadas pela extractão de bauxita, Poços de Caldas, MG**. Rio Claro, Sao Paulo. 2004. 139p. (Doctoral thesis in Biological Sciences)

MORTON, J. B.; BENTIVENGA, S. P.; BEVER, J. D. Discovery, measurement, and interpretation of diversity in arbuscular endomycorrhizal fungi (*Glomales, Zygomycetes*). **Canadian Journal of Botany**, Ottawa, v. 73, n. 1, p. 25-32, Jan. 1995.

NICHOLS, K. A. Characterization of glomalin - a glycoprotein produced by arbuscular mycorrhizal fungi. **Thesis (Ph.D.)** - University of Maryland, College Park, MD, 2003.

NICHOLS, K. A.; WRIGHT, S. F. Carbon and nitrogen in operationally defined soil organic matter pools. **Biology & Fertility of Soils**, Germany, v. 43, n. 2, p. 215-220, 2006.

NOBRE, C. P.; LAZARO, M. L.; SANTO, M. M. E.; PEREIRA, M. G.; BERBARA, R. L. L. Aggregation, glomalin and organic carbon in the Araripe plateau, Ceará, Brazil. **Revista caatinga**, Mossoró, v. 28, p. 138 - 147, 2015.

NOVAIS, C. B.; SOUZA, F. A.; SIQUEIRA, J. O. Phenotypic and molecular characterization of spores of arbuscular mycorrhizal fungi kept in a germplasm bank. **Pesquisa Agropecuâria Brasileira.** Brasilia, v. 45, p. 806-896, 2010.

OEHL, F.; SOUZA, F. A.; SIEVERDING, E. Revision of Scutellospora and description of five new genera and three new families in the arbuscular mycorrhiza forming Glomeromycetes. **Mycotaxon**, v. 106, p. 311-360. 2008.

PAN-BRASIL: National Action Program to Combat Desertification and Mitigate the Effects of Drought. Brasilia: M.M.A. **Secretaria de Recursos Hidricos**, p. 242, 2004.

PATRCIO, E. Communication and education in Mercosur - principles and strategies based on actions to combat desertification and mitigate the effects of drought in the countries of the bloc. **Faculdade 7 de Setembro (FA7)** / Catavento Comunicação e Educaçao / Universidade Federal do Cearà - CE, 2009.

PAULUCIO, V. O. **Chemical and biological quality of an area degraded by clay extraction, revegetated with eucalyptus and legumes inoculated with mycorrhizae**. Universidade Estadual do Norte Fluminense Darcy Ribeiro Campos dos Goytacazes 2007. 106 f. (Doctoral thesis in Plant Production)

PEI, S; FU, H. & WAN, CHANGGUI. Changes in soil properties and vegetation following enclosure and grazing in degraded Alxa desert steppe of Inner Mongolia, China. **Agriculture, Ecosystems and Environmet**, v. 124, p. 1-7, 2007.

PEREZ-MARIN, A. M.; CAVALCANTE, A. M. B.; MEDEIROS, S. S.; TINÔCO, L. B. M.; SALCEDO, I. H. Desertification hotspots in the Brazilian semi-arid region: natural or anthropogenic occurrence? **Strategic Partnerships Magazine**. Brasilia-DF, v. 17, p. 87-106, 2012.

PESSOA, M. F.; GUERRA, A. M. N. M.; MARACAJÂ, P. B.; LIRA, J. F. B.; EDIMAR FILHO, T. D. Study of vegetation cover in caatinga environments with different forms of management in the Moacir Lucena settlement, Apodi, RN. **Revista Caatinga**, v. 21, p. 40-48, 2008.

PINTO, R.M.S.; CARVALHO, V.C.; ALVALÂ, R.C.S. Maps of temporal variability of land use and land cover in the Irauçuba desertification center (CE) for use in meteorological models. In: Brazilian Symposium on Remote Sensing, 14, Natal. **Proceedings** ... INPE, p. 6077-6083, 2009.

POUYU-ROJAS, E.; SIQUEIRA, J. O.; SANTOS, J. G. D. Symbiotic compatibility of arbuscular mycorrhizal fungi with tropical tree species. **Revista Brasileira de Ciência do Solo**, v. 30, n. 03, p. 413-424, 2006.

PURIN, S. **Arbuscular mycorrhizal fungi: Activity, diversity and functional aspects in apple production systems.** Florianópolis, Santa Catarina State University,

147p. (Master's dissertation), 2005.

PURIN, S.; RILLIG, M. C. The arbuscular mycorrhizal fungal protein glomalin: limitations, progress, and a new hypothesis for its function. **Pedobiologia**, v. 51, n. 2, p. 123-130, 2007.

RILLIG, M. C.; MAESTRE, F. T.; LAMIT, L. J. Microsite differences in fungal hyphal length, glomalin, and soil aggregate stability in semiarid Mediterranean steppes. **Soil Biology & Biochemistry**, Oxford, v. 35, n. 9, p. 1257-1260, 2003.

RILLIG, M. C.; MUMMEY, D. L. Mycorrhizas and soil structure. **New Phytologist**, Oxford, v.171, p.41-56, 2006.

RILLIG, M. C.; RAMSEY, P.; MORRIS, S.; PAUL, E. Glomalin, an arbuscular-mycorrhizal fungal soil protein, responds to soil-use change. **Plant and Soil**, The Hague, v. 253, n. 2, p. 293-299, 2003.

RILLIG, M. C.; WRIGHT, S. F.; NICHOLS, K. A.; SCHMIDT, W. F.; TORN, M. S. Large contribution of arbuscular mycorrhizal fungi to soil carbon pools in tropical forest soils. **Plant and Soil**, The Hague, v. 233, n. 2, p. 167-177, 2001.

RILLIG, M.C. Arbuscular mycorrhizae, glomalin, and soil aggregation. **Canadian Journal of Soil Science** v.84, p. 355-363, 2004.

ROSSI, C. Q.; POLIDORO, J. C. Nutrient Cycles and their Relation to Plant Nutrition. IN: TAVARES, Silvio Roberto de Lucena. Course on the recovery of degraded areas: the vision of soil science in the context of diagnosis, management, monitoring indicators and recovery strategies. Rio de Janeiro: Embrapa Solos, 228 p. (**Embrapa Solos. Documentos, 103**), 2008.

SA, I. B.; CUNHA, T. J. F.; TEIXEIRA, A. H. C.; ANGELOTTI, F.; DRUMOND, M. A. **Desertification in the Brazilian Semi-Arid**. Fortaleza - Cearà, Brazil. 2ª International Conference: Climate, Sustainability and Development in Semi-Arid Regions ICID+18, August 16-20, 2010.

SALOMON, J.A.; SCHAEFER, M.; ALPHEI, J.; SCHMID, B.; SCHEU, S. Effects of plant diversity on collembola in an experimental grassland ecosystem. **Oikos**. v. 106, p. 51-60, 2004.

SANTOS, P. R. **Soil attributes as a function of different uses in an irrigated perimeter in the sertâo of Pernambuco**. Federal Rural University of Pernambuco, Recife, 2010. 113p. (Doctoral thesis in Soil Sciences)

SCHINDLER, F. A.; MERCER, E. J.; RICE, J. A. Chemical characteristics of glomalin-related soil protein (GRSP) extracted from soils of varying organic matter content. **Soil Biology and Biochemistry**, 39, p. 320-329, 2007.

SCHNEIDR, J.; KLUBEG FILHO, O.; FONTOURA, S. M. V.; ALVES, M. V. Influence of different management systems and liming in a long-term experiment on arbuscular mycorrhizal fungi. **Ciência agrotecnologia**. Lavras. 35, p.701-9, 2011.

SCHREINER, R. P. Effects of native and nonnative arbuscular mycorrhizal fungi on growth and nutrient uptake of 'Pinot noir' (*Vitis vinifera* L.) in two soils with contrasting levels of phosphorus. **Applied Soil Ecology**, v. 36, p. 205-215, 2007.

SILVA, C. F. et al. Activity of arbuscular mycorrhizal fungi in coastal dunes impacted by mining. In: Brazilian meeting on soil fertility and plant nutrition, 29; Brazilian meeting on mycorrhizae, 13; Brazilian symposium on soil microbiology, 11; Brazilian meeting on soil biology, 8, 2010, Guarapari. Nutrient sources and agricultural production: shaping the future: **Anais...** Viçosa: Brazilian Society of Soil Science, 2010.

SILVA, C. F. et al. Activity of arbuscular mycorrhizal fungi in coastal dunes impacted by mining. In: Brazilian meeting on soil fertility and plant nutrition, 29; Brazilian meeting on mycorrhizae, 13; Brazilian symposium on soil microbiology, 11; Brazilian meeting on soil biology, 8, 2010, Guarapari. Nutrient sources and agricultural production: shaping the future: **Proceedings...** Viçosa: Brazilian Society of Soil Science, 2010.

SIQUEIRA, J. O.; KLAUBERG FILHO, O. Arbuscular mycorrhizae: Brazilian research in perspective. In: NOVAIS, R. F. de; ALVAREZ, V. H.; SCHAEFER, C. E. G. R. (Ed.). **Topics in soil science**. Viçosa: UFV, v. 1, p. 235-264, 2000.

SIQUEIRA, J.O.; SOARES, C. R. F. S.; SILVA, C.A. Matéria orgânica em solos de áreas degradadas. In: SANTOS, G. A.; SILVA, L.S.;

CABEKKAS, L. P.; CAMARGO, F. A. O.(ed.). **Fundamentals of Soil Organic Matter: tropical and subtropical ecosystems**. 2ª ed. Gràfica Metrópole. Porto Alegre, p. 495-520, 2008.

SOUSA, C. S.; MENEZES, R. S. C.; SAMPAIO, E. V. D. S. B.; LIMA, F. S. Glomalin: characteristics, production, limitations and contribution to soils. *Semina:* **Agricultural Sciences**. v. 336, p. 3033-3044, 2012.

SOUSA, F. P. Soil degradation by agropastoral activities in areas under desertification: the case of Irauçuba, Ceará. 89 f.; ill. color. enc. 2009. (Master's dissertation in soils and plant nutrition)

SOUZA, F. A. et al. Classification and taxonomy of arbuscular mycorrhizal fungi and their diversity and occurrence in Brazil. *In*: SIQUEIRA, J. O. et al. **Mycorrhizae**: 30 years of research in Brazil. Lavras: Editora UFLA, p. 15-73, 2010.

STEINBERG, P. D.; RILLIG, M. C. Differential decomposition of arbuscular mycorrhizal fungal hyphae and glomalin. **Soil Biology & Biochemistry**, Oxford, v. 35, p. 191-194, 2003.

SYLVIA, D. M.; WILLIAMS, S. E. Vesicular-arbuscular mycorrhizae and environmental stress. In: BETHLENFALVAY, G. J.; LINDERMAN, R. G. (Ed.). **Mycorrhizae in sustainable agriculture.** Madison: ASA Special Publication, p. 101-124, 1992.

TREVISAN, R.; MATTOS, M.L.T. & HERTER, F.G. Microbial activity in a dystrophic red-yellow Argissolo covered with black oats (*Avena* sp.) in the fall, in a peach orchard. **Ciência Rural**, v. 7, p. 83-89, 2002.

UNITED NATIONS. **Agenda 21**. Senado Federal, BsB, 1996. 591p.

VALONE, T. J.; SAUTER, P. Effects of long-term cattle enclosure on vegetation and rodents at a desertified arid grassland site. **Journal of Arid Environments**, v. 61, p. 167-170. 2005.

VALLADARES, G.S., GOMES, E.G., MELLO, J.C.C.B.S., PEREIRA, M.G., ANJOS, L.H.C., EBELING, A.G., BENITES, V.M. Principal component analysis and ordinal multicriteria methods in the study of Organosols and related soils. **Revista Brasileira de Ciência do Solo**, v. 32, p. 285-296, 2008.

VASCONCELOS S. J. Processos de desertificaçâo no Nordeste do Brasil: sua genese e sua contença. Recife: **SEMA/SUDENE**, 1982.

VEENENDAAL, E. M. Adaptive strategies of grasses in a semi-arid savannah in Botswana. Amsterdam: Vrije Universiteit van Amsterdam, **Ph. D. Thesis**, 1991.170p.

VEZZANI, F. M.; MIELNICZUK, J. Uma Visâo sobre Qualidade do Solo. **Revista Brasileira de Ciência do Solo**, v. 33, p. 743-755, 2009.

WALKER, C.; VESTBERG, M. Synonymy among the arbuscular mycorrhizal fungi: *Glomus claroideum, G. maculosum, G. multisubstensum* and *G. fistulosum*. **Annals of Botany**, v. 82, p. 601-624, 1998.

WARREN, A. Desertification as a Global Environmental Issue**. GeoJournal**. v. 31, p. 11-14, 1993.

WRIGHT, S. F.; ANDERSON, R. L. Aggregate stability and glomalin in alternative crop rotations for the central Great Plains. **Biology & Fertility of Soils**, Germany, v. 31, p. 249-253, 2000.

WRIGHT, S. F.; FRANKE-SNYDER, M.; MORTON, J. B.; UPADHYAYA, A. Time-course study and partial characterization of a protein on hyphae of arbuscular mycorrhizal fungi during active colonization of roots. **Plant and Soil**, The Hague, v. 181, n. 2, p. 193-203, 1996.

WRIGHT, S. F.; GREEN, V. S.; CAVIGELLI, M. A. Glomalin in aggregate size classes from three different farming systems. **Soil & Tillage Research**, Netherlands, v. 94, p. 546-549, 2007.

WRIGHT, S. F.; NICHOLS, K. A.; SCHMIDT, W. F. Comparison of efficiency of three extractants to solubilize glomalin on hyphae and in soil. **Chemosphere**, v. 64, p. 1219-1224. 2006.

WRIGHT, S. F.; UPADHYAYA, A. A survey of soils for aggregate stability and glomalin, a glycoprotein produced by hyphae of arbuscular mycorrhizal fungi. **Plant and Soil**, The Hague, v. 198, p. 97-107, 1998.

WRIGHT, S.; UPADHYAYA, A. A. Extraction of an abundant and unusual protein from soil and comparison with hyphal protein of arbuscular mycorrhizal fungi. **Soil Science**, Baltimore, v. 161, p. 575-586, 1996.

WRIGHT, S.F.; UPADHYAYA, A. A survey of soils for aggregate stability and glomalin, a glycoprotein produced by hyphae of arbuscular mycorrhizal fungi. **Plant and Soil**, v. 198, p. 97-107, 1998.

WU, Q. S; NI, Q. D; QUE, Y. C; HUANG. W. Calcium and Calmodulin Involve in Mycorrhizal and Root Development in Trifoliate Orange Colonized by *Rhizophagus intraradices*. **Not Bot Horti Agrobo**, v. 42, p. 380-385. 2014.

ZHAO, H. L; ZHOU, R. L.; SU, Y. Z; ZHANG, H; ZHAO, L. Y. & DRAKE, S. Shrub facilitation of desert land restoration in the Horqin Sand Land of Inner Mongolia. **Ecological Engineering.** v. 31, p. 1-8, 2007.

ZUQUETTE, L. V; RODRIGUES, V. G. S; PEJON, O. J. Recovering degraded areas. **Revista Engenharia Ambiental**: Conceitos, Tecnologia e Gestào. v, 13, p. 589-619, 2013.

MIX
Papier aus verantwortungsvollen Quellen
Paper from responsible sources
FSC® C105338

Printed by Books on Demand GmbH, Norderstedt / Germany